[Document No. .]

IN THE ASSEMBLY.] [SESSION OF 1855.

REPORT OF COMMITTEE

ON

INTERNAL IMPROVEMENTS,

ON

ASSEMBLY BILL, NO. 16.

B. B. REDDING, STATE PRINTER.

REPORT OF COMMITTEE

ON

INTERNAL IMPROVEMENTS.

The Committee to whom was referred a Bill for An Act to construct a Wagon Road from the Sacramento Valley to the Eastern Boundary of the State, having had the same under consideration, beg leave to report as follows :

That they have had much difficulty in arriving at anything like satisfactory conclusions of the manner of obtaining the end sought. All see the necessity of such a road, and but few doubt its practicability. The immigration to the Pacific demands it ; our progress and prosperity as a State demands it ; the exorbitant price of passage on the ocean steamers demands it ; our connection with the rest of the Union, our safety in war and subsistence in dearth, demands it. Yea, everything demands it, and demands it now ; but how to get it is the difficulty.

In the first place, we have not that unanimity in our own councils that should prevail. Men are differing. The various sections of the country are advocating not only different modes of constructing the road, but different routes upon which to construct it. The people of the north desire a northern route ; the center, a central route ; and the south, a southern route. Some think that private enterprise, with a little assistance from the State, would build a toll road ; others think the State should construct it entire, and let it be free.

All these difficulties can be easily overcome, we apprehend, except the one as to the most convenient practicable route.

Our State is so new and unsettled, its mountain gorges and cañons are so little explored, and its waste places and deserts so little known, that your Committee have not been able to agree unanimously upon any particular route.

The first route which occupied our attention was the one called Noble's Pass. Those who have traveled it, represent it as being a very excellent pass, particularly for summer travel. The distance through that pass from the Sacramento Valley to the Humboldt is about two hundred and fifty miles. The ascent of the Sierra Nevada either way is very gradual, and requires little or no grading. There is an abundance of grass on this route for the stock of a large immigration. Water is not so abundant, there being frequent scopes of country from twelve to eighteen miles in which there is a scarcity of water. The highth of this pass is represented to be between four and five thousand feet above the level of the sea.

The objections to this route are, that it is too far north ; that the immigration will not wish to take it, because it is too far north, and much further than the great majority will like to travel to reach their destination.

We have been told that a Mr. Jenkins proposes to make a good wagon road over that route, for the sum of ten thousand dollars. Major Corponing, United States Mail Agent at Salt Lake, says that it will take a much larger amount, and that it will then be impassable for a great portion of the year on account of snow.

The next route coming south which occupied our attention, is called Beckwith's.

This route will leave Marysville and go by Bidwell and the American Valley into Beckwith's Valley ; or leaving the same place, will go by Gibsonville through Beckwith's Valley, down the Truckee, across the desert to the Humboldt. Stages run daily from Marysville to Bidwell, a distance of thirty miles, and of course the road will need but little work or outlay to make it a good wagon road that far. From Bidwell to the American Valley, a distance of about seventy miles, there is now what is called a good mountain road, but will need much improvement to make it such a road as is contemplated by the Bill. Over this portion of the way, it is said private enterprise will run a line of stages next spring. From the American Valley the road goes to Beckwith's Valley, the distance of seventy or seventy-five miles. This portion of the road is better than that lying between the American Valley and Bidwell, but will need much improvement.

The other route leaving Marysville is about as follows : To Gibsonville, a distance of sixty-two or sixty-three miles, the road is now traveled with stages. This would not need any work to make a good wagon road ; at least, no more than the counties through which it passes could afford to do to make it good. Along this portion of the route there is plenty of water. From Gibsonville this way, for fifteen miles, there is no grass, the country being covered with mountain chaparrel. At Gibsonville, grass sufficient can be obtained. From Gibsonville to Jameson Creek at "76," a distance of fifteen miles, the road passes over high, stony, chaparrel ridges, nearly destitute of food for animals. This portion will require an outlay of from $10,000 to $15,000 to make it good. From "76" to the Mohawk Ranch, a distance of twelve miles, the road is now good. From the Mohawk Ranch to Beckwith Valley, a distance of twelve miles, the road is tolerable good,—will need some work. From Beckwith Valley to the Humboldt, a distance of about eighty miles, it is said a good road can be made at a moderate expense. The information as to this part of the route is not very satisfactory. The writer of this came up the Truckee in 1849, and found it anything but a good road, having crossed that stream twenty-seven times in about twenty-seven miles. It is said, however, that by new routes, crossing the stream so often can be avoided—that it will only be crossed from three to seven times. We suspect that this portion of the route will need a great deal of labor to make it a good wagon road. The whole distance from Marysville to the Humboldt is two hundred miles, more or less.

The next route is the one up the South Fork of the American River, through Johnson's Cut-Off and Hope Valley into Carson Valley.

From Sacramento, by this route, the distance is one hundred and ten miles. From Sacramento to Placerville, some fifty miles, there is now a good road, over which stages pass daily, making the trip in five or six hours. From Placerville, sixteen miles, the road is good, and needs but little work. From this place along the banks of the South Fork to Slippery Ford, a distance of twenty-five miles, the road will need considerable work, owing to points or ledges of rock running in close to the river. From Slippery Ford to the summit of the Sierra, is only six miles and two hundred yards, by actual measurement. At Slippery Ford there is quite an obstacle in the shape of a rocky hill ; it can be avoided, however, by keeping up the South Fork, which will need bridging. There being fine water power and excellent timber on the spot, the bridging can be done comparatively cheap. The ascent from Slippery Ford to the summit is most beautiful and gradual. The summit is 8,300 feet above the sea. From the summit to Lake Bigler Valley is some three miles, and by side-hill grading can be made a good road for wagons and stages. Here, however, is the greatest difficulty on this route. Lake Bigler Valley is 7,150 feet above the level of the sea, leaving the difference between the valley and the summit of the Sierra of 1,150 feet, which has to be descended in three miles, making a grade of some three hundred and eighty feet to the mile. From the eastern foot of the Sierra, the road turns south-easterly across the southern part of Lake Valley into Hope Valley, a distance of five miles. Hope Valley is 7,540 feet above the sea. From the eastern foot of the Sierra to the eastern part of Hope Valley the road is good. From Hope Valley through the cañon to Cary's Mill, is six miles. Cary's Mill is 6,240 feet above the sea, making the fall in six miles, 1,300 feet. The cañon has been worked some, but needs much more. The work is of that character which is easily to be accomplished. The people of Carson Valley would do this, it is said, without charge to the State of California. The total distance by measurement from Placerville, in El Dorado County, to Carson Valley, is fifty-nine and a half miles,—seven miles only of which is covered with snow so as to hinder traveling. This route passes through a very beautiful and picturesque country, well watered and timbered.

Lake Bigler is a beautiful sheet of water, from eight to twelve miles wide and about forty long, lying about 7,500 feet above the sea, between two lofty ranges of the Sierra. The valley surrounding it is traversed by many lovely streams, and in the summer is covered with an exuberant growth of grass. Carson Valley is from ten to fifteen miles wide and from thirty-five to forty miles long, and is now settled by farmers, millers and miners. The soil is rich, and well adapted to the production of grains, grass, etc. Water for irrigation and milling, is abundant. This bids fair to be a thickly settled country, and must necessarily have communication with our State. We have described this route more particularly than any other, for the reason, our information has been greater and more reliable.

The advantages of this route, are—

First. It is a central one.

Second. It is much the shortest route to the Sacramento and San Joaquin valleys.

Third. The greater part of the way can now be traveled and needs but little work.

Fourth. The snow is bad, only for about seven miles.

Fifth. It connects us with the Carson Valley, a valley which will soon be densely settled, and will be the Half-Way House from the Great Salt Lake country.

Sixth. Because by recent explorations, it is found by continuing this route, a good wagon road can be made, with but little labor, north of the Great Salt Lake to the Mormon settlements, one hundred and fifty miles nearer than any now traveled, making the total distance from Sacramento City to Salt Lake, not more than five hundred and fifty miles.

The estimated cost of this route, is from $30,000 to $75,000.

Reference is made to the report of Mr. Henderson, County Surveyor of El Dorado Co., marked "Exhibit A." and hereto attached.

There is also another central route, to which our attention has been called; leaving Sacramento and passing by Diamond Springs, El Dorado County, through

Sly Park, on or near the old Carson Route, through the cañon into Carson Valley.

This route has been lately explored by Major Case and others, who represent it as being an excellent one for a wagon road. Others say that it is the old Carson Route, and that it can never be made as good a road as the one through Johnson's Cut-Off.

By this route, from Diamond Springs to Carson Valley, the distance is sixty-seven and a half miles according to the calculations of Major Case; which distance may be shortened some eight or ten miles, by cut-offs.

Upon careful examination it may be found, that the route on or near the old Carson Road, by Diamond Springs through Sly Park, is the best route we have named, as Major Case says that the road from Diamond Springs to Carson Valley, can be shortened to fifty-five and a half miles, and that but twenty-five miles will require work. The Committee have no estimate of the cost of this route.

See Major Case's report marked "Exhibt B."

The fifth route is the southern route, leaving Salt Lake and running south through the southern settlements of the Mormons, along the route of the proposed Military Road of the General Government, to the eastern boundary of this State; thence through the Cajon Pass and San Bernardino, to Los Angeles; thence, up the Coast Road to the central portion of the State, or from Los Angeles through the Tejon Pass, through the Kern River country, by the Sink of White River, across Tule River into the King's River and Four Creeks' District; thence down the San Joaquin to Stockton.

This route passes through some most beautiful and fertile country, on the Four Creeks and King's River, the whole of which is public land and is a most excellent place for settlers. The soil is well adapted to grains of all kinds and grass, and is also well suited for stock raising. Water and timber are abundant. In order to understand this route more definitely, we propose to describe it in three sections, commencing at Stockton.

From Stockton to the Tejon Pass, a distance of two hundred and ninety miles, the road crosses many streams and passes over a lovely country, known as before said, as the Four Creek Valley. This section of the road will cost from $10,000 to $12,000. The Tejon Pass is about 4,000 feet above the sea and about 2,000 feet above the Tulare Valley. The grade is said to be very easy, indeed a good road now exists over it. From Tejon Pass to Los Angeles, is ninety miles; from Los Angeles to the town of San Bernardino, is fifty-six miles. This section will need some considerable work in the mountains near the Tejon Pass. The Los Angeles people have already spent about $6,000 on this road. It is said that $6,000 or $8,000 more, would make a good wagon road of it. Along it, water and grass are abundant and the soil good. From Los Angeles to San Bernardino, there is now an excellent road. From San Bernardino to the eastern boundary of the State, the distance is about two hundred and fifty miles, and forms the third section.

The Cajon Pass is some twelve miles from San Bernardino. This pass will need much work and an outlay of $20,000 more or less. The distance through the Cajon Pass is some twelve miles; once through it, you are in the Great Salt Lake Basin with a good road to Salt Lake City. There are two *jornadas* on this route, each about fifty miles long. Wells would have to be dug and watering places fixed, or the immigration would suffer intensely. Grass is scarce in many places. The great advantages of this route are—

First. There are no high and abrupt mountains to pass; and,

Second. The snow is no impediment, consequently it can be traveled at all seasons of the year.

The great objection to it is, that it is too circuitous and nearly, if not quite, as far again to the central portions of the State, as the route by Carson Valley to the Sacramento Valley.

The distance from Salt Lake *via* Los Angeles, Tejon and King's River to Stockton is about 1,100 miles.

Your Committee would say, that from all the information in their possession, and from a careful consideration of the advantages and disadvantages of all the routes, they feel constrained to recommend the location and construction of the road upon the central route, through Johnson's Cut-Off.

Your Committee would further say, that they have not been able to procure anything like a definite estimate of the cost of such a road. That the estimates which those who have traveled over these various routes are, as we have before stated, from $30,000 to $75,000. That the cost can only be ascertained by actual survey and calculation.

Your Committee would further report, that they have made many alterations and changes in the bill introduced by Mr. Farwell. That they have attempted to simplify the same, and to require, if possible, the construction of the proposed road in time to relieve the immigration of the present season; and having so amended said bill, they report the same back and recommend its passage.

C. T. RYLAND,
Chairman of Committee on Internal Improvements.

[EXHIBIT A.]

Placerville City,
February 23, 1855.

Dear Sir:

Yours of the 13th of November last, requesting me to give you some information concerning a pass through the Sierra Nevada, in this county, has been received. At that time I had never crossed the Sierra within the limits of this county, and not being in possession of any reliable information on that subject, I thought it improper to say anything on the subject at that time.

Since that, I have been called upon by the citizens of this place to examine for a road route from this to Carson Valley; and, accordingly, on the 31st of January last, with the necessary outfit, I left for the aforesaid purpose, and arrived at the Mountain Ranch, a distance of thirteen miles, near sundown. Between Placerville and this place, there is a good wagon road at present. After leaving the Mountain Ranch a quarter of a mile, we leave the old road and turn to the left to avoid a hill; we then descend, intersecting Johnson's road, sixteen miles; then keeping along the same about three-quarters of a mile, where it turns short to the left to cross the bridge, but we keep straight forward, on a beautiful flat, and continuing on good ground for a road, until we arrive at eighteen and a quarter miles, then turning to the right, up a cañon, and on a ledge of rocks, and at nineteen miles, cross the river; (from the point that we leave Johnson's Road to this is nearly level.) Crossing the river here, we land on good ground for a road, and for near half a mile will require but little more to make a good road than to clear away the timber. There are, however, occasionally a rock bluff to encounter. We cross the river at the aforesaid point for three reasons; first, we need not descend from the time we leave Johnson's Road. Second, to avoid snow on the south side of the river. Third, to have the advantage of several flats or benches that present themselves.

At twenty-eight miles we passed what is called the foot of Peavine Hill. At twenty-nine and a half miles pass Sugarloaf Rock. At thirty-three and a half miles pass Cottage Rock. At forty-one miles arrive at Slippery Ford on Saturday evening, the 3d instant.

Thus far a good road may be had by excavating rock in spots along the line.

Leaving Slippery Ford and running up and onthe north side of the South Fork there will be rock to contend with for half a mile, then gradually ascending at about three and a quarter degrees through a beautiful timbered flat, to the distance of seven and a half miles from Slippery Ford to the summit.

In going out, we went the present trail, and arrived at the summit at a distance of six miles and sixteen chains from the aforesaid ford, then descended the old trail into Lake Valley.

This pass is 8,308 feet above the sea, and the valley at this point is 7,150 feet above the sea. The axes of this valley is nearly north and south. From this we bore in a north-easterly direction along the old trail for the distance of eleven miles. Here we halted for refreshments, and while the cook was preparing a lunch, Messrs. Taylor, Silman and myself, went in a west by north direction for near half a mile, when we found ourselves on the east bank of Lake Bigler. This lake at this point is eighty-six miles wide east and west, and said to be fifty miles north and south.

After refreshments, we proceeded on our journey, by leaving the wagon road at this point, and running north seventy-five degrees, and east up a mule trail three miles and twenty-six chains brought us to the east summit in the pass 8,320 feet above the sea.

Thence south sixty-four degrees, east three miles and four chains brought us to Dr. Daggett's, at the foot of the mountain, and on the west side of Carson Valley, making the whole distance from Placerville to this place sixty-five and three-eighths miles.

The valley at this point is 5,200 feet above the sea.

Coming down this last mountain, we are obliged to descend over one thousand feet per mile. I therefore consider this last part of the route impracticable.

This is a beautiful valley, and is said to be fifteen miles wide at this point. The Mormon station is about two miles north of this. We then proceeded up the valley to Mr. Cary's mill, a distance of fourteen miles.

In passing along the western border of this beautiful valley, almost every quarter of a mile we were delighted by crossing a beautiful cool crystal stream of water, descending from the white-capped mountain peaks, to irrigate the fields below.

Mr. Cary's house, near the upper end of Carson Valley, is one mile below what is called the mouth of the cañon, is 6,240 feet above the sea. Then beginning at the bridge called the mouth of the cañon, and running up the cañon five miles, brought us into Hope Valley.

This valley is 7,540 feet above the sea. From Carson Valley to this, the road need not have an elevation of over three degrees at any place. Leaving Hope Valley, we turn north sixty-five degrees west, gradually ascending the hill with five degrees elevation to gain the pass between Hope and Lake valleys; at six miles and forty-six chains reached the pass, 8,240 feet above the sea.

Then descending gently over ground that will require but little more labor to make a good road than to clear the timber, into Lake Valley, which we reached at ten miles.

In leaving Lake Valley the west ridge may be ascended in two miles, with an elevation of six and a quarter degrees. Twelve miles brought us on to the west summit; at thirteen miles and seven chains reached the old trail at a point forty-six and a half miles from Placerville to the mouth of the cañon at the head of Carson Valley fifty-nine and a half miles and seven chains.

While on the above trip, I took observations every night, and sometimes at noon for latitude; many of them taken amid wind and clouds. The instrument used was a theodolite, so that great accuracy cannot be expected.

Having no barometer, I was lucky in obtaining a large thermometer, and with it by the boiling point of water, obtained the above hights.

The longitude laid down on this map is Eddy's.

There are other passes through the Sierra suggested to me, one some four or five miles south of the one we came through, and another five or six miles north, which I think it would be well to examine before a final location is made.

I am inclined to believe the route here laid down, is the most direct that can be got, and that it will be the least liable to be shut with snow, of any other for many miles south.

In those passes or summits that we went over, the snow was about one foot and a half deep, and the first thirty-six miles the ground was entirely bare.

The reason I recommend an examination of those other passes is, that from the time we reach the river, there are occasional bluffs of rock to encounter which might be less on some of the other routes.

I am, Sir,
Your most obedient servant,

WM. HENDERSON,
County Surveyor El Dorado County.

[EXHIBIT B.]

MEMORANDA OF OBSERVATIONS ON ROUTE FOR WAGON ROAD FROM SACRAMENTO TO CARSON VALLEY, FEB. 16 TO FEB. 23, 1855.

The road from Sacramento to Diamond Springs, being well known, needs no description—distance fifty miles.

Diamond Springs to Graham & Slaughter's Ranch, in Pleasant Valley—distance nine and a quarter miles; road in good condition for staging, and of a level character. Some good grades have been made where hills were of much natural acclivity. This road passes through a valley well settled and in good cultivation. Several ditches for supplying miners with water pass through this valley. Graham & Slaughter have a fine large rancho, and raise large quantities of grain and fodder. There is a good public house at this point.

Graham & Slaughter's Ranch to Sly Park—distance six and a quarter miles. The road still continues good, and much mining is done for a distance of about four miles. Country well watered, and plenty of timber and wood; one and a half miles leave Clear Creek, which comes down from north-east, we turn east up through Dry Creek for a short distance. The road has been graded round a long hill, which makes it easy of ascent, the increased altitude in the advance being so gradual that one hardly perceives he is really rising thirty feet to the mile, especially through those portions which have been worked. From a point three miles from Graham's a very material improvement can be made by a cut-off which will avoid an ascent and descent of one hundred feet, while it shortens the distance from the point named. The road is the natural one; and I will here remark that all roads leading into California have been made upon this principle. A train must follow a flat route instead of a sideling one; in descending a ridge, to do so at right angles, instead of risking upsets by a gradual or side-hill slope. We here see the last mining on this route; the country assumes more the mountain peculiarities—pines more frequent, some firs, less oak. We reach Sly Park by an easy descent, and come upon one of the finest mountain valleys imaginable, entirely surrounded by high ridges, except the entrance, and exit openings of the Sly Park Creek, which runs upon the southern side of the valley or park. This park is a perfect level of about one hundred acres of fine land. Large pines are growing upon it, which give it the appearance of an old sit-place, but much more beautiful than art could have made it. The creek is filled with trout of most delicious flavor, and of which we had a taste. This will eventually become one of the places of summer retreat in California.

From Sly Park to Iron Mountain the distance by road is nine and a half miles.

Leaving Sly Park we cross the creek, and by a road made by the proprietor, Mr. Stonebreaker, we wind around a hill known as Stonebreaker's Hill, rising at an average grade of say one to eight and a half feet for one mile—then up a a gentle trotting ascent we pass on through a finely-timbered country towards Iron Mountain. This Stonebreaker's Hill is the western terminus of a sierra, or ridge, of the grand chain of the Sierra Nevadas. It is a high ridge crowning upon the top, and in general feature resembling the back of a fat hog. The road, after leaving the graded part, is one and a half miles from Sly Point, running along the spine. This back varies in width from one hundred feet to one hundred yards, and even sometimes more, and exhibiting the most splendid growth of timber. This is a natural road-way, and in its approach to the main ridge, is steadily and almost imperceptibly to the traveler attaining an increase of altitude of nearly one hundred feet to the mile. Some four miles from Sly Park we see the old Camp Creek turn-off, where a diversion is made to the right for a circuit of two and a half miles to the camping and watering ground of Camp Creek. Coming in again some one and a half miles ahead, one mile before we reach Iron Mountain, we stop at one of the finest springs, containing any quantity of clear, pure, cold water.

At Iron Mountain the old Carson River route comes in over a divide running nearly south, and up a pitch of one to four feet for one hundred yards. This old and well-known spot, formed of lava rock, once rough, has by use become so smooth as to be almost impassable for even a sharp-shod mule without a load. From this point the most beautiful and magnificent mountain views are presented at the north. The long array of snow and rocky peaks stand out in line extending north-eastwardly as far as the eye can reach, and so westwardly to the junction of this ridge with the grand Sierra, near Silver Lake, distant nearly twenty miles.

From Iron Mountain to Alder Springs, distant eight and a half miles, the grade still continues easy, and the appearance of the country the same, with rather more firs. We have now on our right hand the head-waters of the Cosumnes—on the left the head-waters of the South Fork of the American River.

From Alder Springs to Leek Springs, two miles. Alder Springs is at the foot of Alder Hill. This hill commences with an ascent of about one to five feet, and drops away in a mile from one to eight—then a quarter of a mile nearly level—then descends from one to ten to Leek Springs. This is now the hardest part of the road, but a cut-off a short distance below the foot of Alder Hill, as is proposed, to the foot of Silver Lake, will avoid all this, and materially shorten the distance.

From Leek Springs to Independence Flat, three and a half miles; Volcano Fork two and a half, and Tragedy Springs four miles. At Leek Springs the Grizzly Flat trail comes in through the valley, which extends southerly from the road, near to which are the Springs. This is a good camping-ground, and in the immigration season must be an excellent place to camp; but at this season it is rather cool, requiring frequent additions of fuel to the camp-fire, and some shaking, to keep warm, as some of our party can affirm. The thermometer at sundown, eighteen degrees—at sunrise, ten degrees. The springs were entirely buried in snow, and some "last immigrant" had, in a spirit of wantonness, burned down the only shanty on the spot, and we had to sleep on the snow. From this place to Independence Flat the road is ascending and still hard, one and a half miles to a turn to the northward of the ridge we have followed from Sly Park—the road still keeping the spine and on to Independence Flat. All that can be said of this flat is that the back is a little broader here than above. Still on two and a half miles we find the Volcano Road branching off more southerly than the Grizzly Flat, or main road; thence two miles to a turn up a rocky hill—then bending still more, descending, we reach Tragedy Springs in two miles. The country now all around has a rough mountain character.

Nearly north-east from Tragedy Springs is the Silver Lake, with the valley extending southwardly, and the old road goes round the south end of the valley, and, by a circuitous route, reaches the point of crossing in twelve miles; but Messrs. Ruse & Barnard made a cut-off around the northwardly end, below the lake, and reach the same point on the summit in seven and a half miles. This lake is the head-water of the South Fork of the American River. Here is plenty of water, and a trading-post in summer. From the summit, (the eastern,) which is a rocky hill and will require much labor to make easy, two and a half miles brings to Clear Lake, sometimes called Big Red.

A near route is described through a cañon or ravine a little north of the summit from Silver Lake, which might be bridged, or have rocks removed so as to form an easier passage for wagons.

Big Red, or Clear Lake, is situated between the two sierras, or summits, western and eastern; from it to the eastern summit, three miles, not very hard; thence easy descent three-quarters of a mile to Little Red Lake, in Hope Valley; thence four miles, through Hope Valley, to the entrance to the cañon leading into Carson Valley, four miles; thence twelve miles to Mormon Station.

The theory of this proposed route is one grand central route, into which the Volcano and Grizzly Flat roads on the south, and the Placerville and Georgetown roads on the north, naturally fall, before we come to that point where the largest amount of expenditure is required to form a good wagon road, thus giving them the benefit of the greater part of the outlay upon the main trunk of the road, which will enable the State to also aid the roads to the extreme north and south portions of the State.

The cut-off from near Alder Hill has been mentioned. This is one of the most important. Others, of less importance, may also be easily made, making, in the aggregate, say twelve miles to be deducted from the distance to Carson Valley.

Thus, from Diamond Springs, by the present Sly Park route, sixty-seven and a half miles, from which deducting twelve, we have only fifty-five and a half miles, and only about twenty-five miles which require more than a trifling expense.

The party making these observations consisted of Major Case, F. Tukey, Esq., C. G. Scott, Major Graham and W. Stonebreaker, by whom the matter is respectfully submitted, in the hope that the importance of so central a route will not be overlooked.

[EXHIBIT C.]

SACRAMENTO,
January 20, 1855.

Sir :

In responding to your several inquiries, relative to the geographical character of the section of country lying between the Sacramento and Humboldt rivers, as situated between the parallels of 40 and 41° north latitude, and the practicability of constructing a National Railroad through that district, the following statements are a summary narrative of my views, as deduced from personal observations through that region, in the summer of 1850.

There are four distinct ranges of mountains embraced within the above limits.

The first is the Sacramento Mountains, which form the eastern boundary of the Sacramento Valley north of the Three Buttes, as far as that valley extends.

The second, is the Pitt River Mountains, commencing immediately east of Grizzly Valley, in about latitude 40° 15′ north, and longitude 120° 25′ west, and extends north-westerly to a point where they intersect a ridge running east, and west in latitude 41° 35′ north, and directly south of Rhett Lake.

The third is the Sierra Nevada, which crosses the above district in longitude 120° west, ranging in line with the eastern boundary of California, and varying but little from a due north and south course.

The fourth is the Pillars of Atlas, which lie in the interior basin, along the western border of the Pyramid, and Low Mud Lakes, and the Boiling Spring Valley ; and ranges in a north north-east course to their terminus, near the mouth of the High Rock Cañon in about latitude 41° north.

The North Fork of Feather River does not take its rise in the Sierra Nevada (as is generally supposed) but on the east flank of the Sacramento Range.

This ridge commences near Pitt River, in latitude 41° north, and thence bearing in a south by east course, and parallel with the Sacramento River, extends about one hundred miles to its terminus at the Table Mountain, near Ophir. The summit of this ridge is some 35 or 40 miles distant from the Sacramento River. The western slope is broken into lateral ridges ranging west ; at nearly right angles with the main ridge, and through the ravines between them, flow the several little streams, known as Rapid, Colo, Cloves, Buttle, Antelope, Pine, Dry, Deer, Chico and Butte creeks. These streams issue forth from deep and impassable cañons, their foaming waters roaring loudly, as they rush impetuously through the rocky avenues that bind them. The surface of the ridges is rocky and barren, sparingly studded with dwarf oak and manzanita bushes, excepting in the higher altitudes, where the pine and cedar flourish.

This region, commonly known as the Rocky Desert, presents a dreary and uninviting appearance to the traveler, who will never forget the power of the sun's direct and reflected rays, while passing over them in midsummer.

The character of the eastern side of the Sacramento Mountains, is very different from that of the western—with the exception that it furnishes a corresponding number of little creeks flowing eastward into the valley of the North Fork of Feather River; their several conjunctions, in combination with a few small streams issuing from the Iron Hills, form the principal constituents of that river. The declivities of this side of the mountain are more precipitous, and are densely covered with forests of pine and cedar, and the different varieties of evergreens. The temperature of the atmosphere is many degrees colder, in the same season, and at equal altitudes, than that of the west side.

The more elevated peaks in many places, on this side, and north of the west branch of the North Fork, are covered with snow during the greater part of the year; advancing northward, the snow increases in quantity, and the mountains in altitude, until they abruptly terminate at Sasseus Butte. This is the fountain head, from whence innumerable little streams descend, bounding wildly over the craggy precipices, rushing and roaring down through deep ravines, till they reach the romantic vales below, where they meander and play midst the wild bowers of grapevine and willow, intermingle together and form the North Fork of the Feather River.

The Iron Hills embraces a tract of country that lies between the Sacramento Mountains and the Pitt River Range, and extends from the east branch of the North Fork of the Feather River northward as far as Pitt River, and forms the dividing ridge between those two streams; but from the broken character of the country, no definite line can be drawn, until the culminating points are ascertained by actual surveys.

This region constitutes a series of hills and valleys, with but little variation in their hight, and heavily timbered. The soil is of an argillaceous composition, mixed with disintegrated quartz, and strongly impregnated with red oxide of iron. In various places intervening the hills, are extensive flats of alluvium, their surface richly carpeted with grass and clover, and in many places an abundance of wild strawberries. Also embosomed among them are several small lakes.

The east branch of the North Fork receives most of its waters from these hills, and is the only stream of importance that empties into that fork from the east side.

The Pitt River Mountains are a low ridge of igneous origin, presenting a black and rugged appearance. The main body of them lies about midway between the Sierra Nevada and Sacramento Ranges, crossing the country obliquely. Their summits are destitute of snow throughout the summer months. The Pitt River forces a passage through this ridge at the Nine-Mile Gap, and thence gliding along by the northern base of Camp Hill, crosses the Pitt River Valley through the Great Meadows, and enters the Iron Hills at the point where the Lassen trail leaves that river. Camp Hill is an isolated mount, in the open plain, a few miles south-west from the Nine-Mile Gap. The distance from this gap to the Sierra Nevada is about fifty miles, and the country between it and the South Fork of Pitt River is mostly a level plain, of a sandy character, and entirely destitute of timber. In the southern portion of the Pitt River range, in about Latitude 40° 30′ North, is the passage known as the Black Butte, or Noble's Pass, which was discovered by Captain Hough and party, early in the spring of 1851, while on a prospecting tour from Indian Valley. Immediately to the south of this pass rises the Black Butte, which is the southern continuation of the Pitt River Mountains; and between this southern division and the Sierra Nevada lies Honey Lake Valley.

By some unaccountable mistake in the compiling of Eddy's Official Map of California, the above range of mountains have been confounded with the Sierra Nevada, while the continuation of the real Sierra Nevada from Lassen's Pass southward is cut off, and their true geographical position left blank. Nothing can appear more ridiculous than the idea of Pitt River having its source east of the Sierra Nevada, in the Interior Basin, and thence running eastward through that range of mountains, as there exhibited.

Honey Lake Valley is about twenty-five miles broad, and bounded on the east by the Sierra Nevada, on the west by the Pitt River range, and on the north by a spur of the Sierra Nevada that branches off immediately to the north of the passage through that range known as Fredonyer's Pass, and thence bears westward till it comes in contact with the Pitt River range, immediately to the north of Noble's Pass at the Black Butte. This spur forms the dividing ridge which separates the waters that flow north into Pitt River, east of the Pitt River Mountains, from those that flow south into Honey Lake Valley, and forms an almost impassable barrier along the northern borders of that valley. About twenty miles north of this spur, and high up on the west flank of the Sierra Nevada, inclosed in a little valley, lies Snow Water Lake. A few miles to the west of this lake is Castle Bluffs, bordering the valley of the South Fork of Pitt River, above the Great Bend. This valley varies from one to three miles in width.

The South Fork takes its rise in the Sierra Nevada, at the north side of the spur above alluded to, and runs northward along the western base of the Sierra Nevada till it reaches the Castle Bluffs, where it turns west, forming the Great Bend, and thence north, to its junction with the North Fork of that river.

Honey Lake Valley is one of the most romantic and beautiful in the world. The surrounding flanks of the mountains are densely covered with forests of pine, and its bottoms luxuriantly coated with grass and clover. On the west side of the valley, at the extreme base of the Pitt River Mountains, lies Honey Lake,—its green banks decorated by no monuments of art, and its placid waters ruffled by naught but the elements.

A few miles north-west of the Lake is the Black Butte Pass, from the summit of which there is a direct and nearly level route northward along the western base of the Pitt River Mountains to their terminus, and thence to the Umpqua Valley in Oregon.

At the north-east corner of Honey Lake Valley, is the Great Pass through the Sierra Nevada, in about Latitude 40° 35′ North. This passage, when viewed from the summit of the mountain spur bordering the north side of the valley, presents a grand and sublime appearance, the peaks to the north and south sides of the pass, elevated many thousand feet and covered with perpetual snow, dazzle in the sunbeams, forming a beautiful contrast with the green shade that overspreads the surrounding localities.

In beholding this region, the mind is overwhelmed with awe in contemplating the power of God, as displayed in the stupendous magnitude of this rupture where Nature, in her terrific convulsions, has cleaved the mountain asunder, making the high places low and the crooked straight, whereby her children may pass through in safety.

This pass is about twelve miles long, and nearly as broad as the south pass, through the Rocky Mountains, but much deeper depression ; the ascent, to its culminating point, is very gradual, the grade over the steepest places not exceeding one hundred feet, and the average not over seventy feet to the mile.

The mountain sides adjoining the pass are heavily timbered, especially on the western declivities, and present a formidable barrier to a passage over them, either to the north or south.

Immediately to the east of the pass lies McNamey Valley, which is about sixty miles long, extending from the Sierra Nevada to the Pillars of Atlas.

A succession of low ridges that slope into this valley from the north and south sides, causes a variation in its width from ten to twenty miles.

The McNamey River, a small stream flowing through this valley derives its source from the eastern flanks of the Sierra Nevada, and as it winds its way eastward to Mud Lake, the volume of its water is augmented by several little streams issuing from between the adjoining ridges above alluded to.

These ridges are of a rocky and barren character, entirely destitute of timber, and their borders only covered with a few stunted sage brushes ; the soil is sandy, and the general aspect of the country is sterile and parched with heat ; but the bottoms along the margin of the river are abundantly coated with grass, and show excellent signs of periodical floodings. To the north of the dividing ridge at the head of the Snow Creek Valley, and close by the eastern base of the Sierra Nevada, lies Holloway Lake, confined in a deep valley, about seven miles broad, and bounded on the east side by the Palisade Rocks, which form a perpendicular precipice for several miles, and varies from fifty to one hundred feet in hight, their upper surface spreading out in a horizontal strata, covered with a thin layer of soil, which is baked hard and destitute of vegetation. Holloway Lake is about six miles broad and from fifteen to twenty miles in length, and is fed by several little streams formed by melted snow on the Sierra Nevada, the two principal of which descend, one from the middle gap and the other opposite the southern extremity of the Lake.

Between the lake and the southern end of the valley that confines it are extensive flats covered with a white alkaline incrustation several inches thick, with here and there clusters of wild sage and geesewood bushes. On the west side rises the Sierra Nevada, with its broad flank, broken by successive ridges of granite protruding through a superimposed strata of trappean rocks. The shelves in many places between the protrusions have a gradual slope, and all covered with an alluvial deposit of sufficient thickness to support a vigorous growth of timber.

The Pillars of Atlas, the fourth range of mountains before spoken of, are of volcanic origin, rugged in their appearance, and entirely destitute of timber.

At a point near the north end of Lower Mud Lake, and opposite the Great Boiling Spring, the continuity of the ridge is broken, forming one of the grandest ruptures in Nature. One side of the cleft rises to the giddy hight of a thousand feet, with nearly a perpendicular declivity, while the other inclines off in gradual retiring strata to nearly double that distance. The bottoms through that gorge are on a level with the adjoining plain, and affords a free passage for the McNamey Creek through into Mud Lake. About twelve miles north of this point, and a little west of the Pillars of Atlas, is situated Mount Observation, which rises to an elevation of three thousand feet above the plains of the Interior Basin, and is of a conical shape, and perfectly barren.

The physical character of the surrounding localities, when viewed from the summit of this mountain, presents a dark and uneven surface, composed of several basaltic ridges, ranging from north to south, where they slope off into the McNamey Valley. These ridges are broken by deep and yawning chasms. The surface in some places above their brinks spreads out in inclined plains, covered with fragments of lava so completely imbedded as to form a solid pavement. In other places it rises up to inaccessible pinnacles, evidently elevated in a state of fusion. In fact, so plainly are the effects of volcanic action exhibited throughout these localities, that they forcibly impress the mind that the doors of the infernal regions have here been but recently closed, while a dead silence seems to brood over the whole district, presenting a scene of gloom and desolation scarcely ever equaled.

Directing the view beyond this burnt district to the west, the gigantic Sierra Nevada is seen ranging north and south, with its snow-capped summits rising in majestic grandeur as far as the vision extends. Lassen's Pass in a north-west, and the Great Pass in a west south-west direction, are distinctly seen—likewise the Middle Gap between Snow Water and Holloway lakes, about sixty-five miles distant from this point.

On the north-west side of Mount Observation lies Crater Valley, which is the aperture of an extinct volcano of an oval shape, and about five hundred yards in diameter. At the south-west side of the crater is situated Ladder Cañon, which is utterly impassable. Beyond a rocky defile at the north side of the mount rises a high ridge of basaltic cliffs, presenting a bold and precipitous front to the eastward. At the foot of these cliffs lies Cañon Valley, bounded on the east side by the Pillars of Atlas, and extending in a north-west course about eighteen miles to the termination of that ridge. A few miles from the head of the valley is the Augitie Causeway. This is a volcanic fissure of about a mile and a half long, and from one to two hundred feet broad, each side forming a perpendicular wall fifteen feet high. The bottom between them has been filled up with drift to an unknown depth.

The valley above and below this causeway is entirely destitute of timber ; the soil is sandy and covered with innumerable little pieces of obsidian of different colors. Near the northern end of the valley is situated the mouth of High Rock Cañon ; this is a deep fissure passing through a tabular ridge of basalt and greenstone rocks, its width varying from twenty-five to fifty yards. The walls lining its sides are perpendicular, and several hundred feet in hight. A small creek issues from the mouth and discharges into Cañon Lake, which is a small body of water that lies on the opposite side of the valley. The length of this cañon I am unable to give, having been in only a short distance. Immediately to the east of the north division of the Pillars of Atlas, lies Boiling Spring Valley, bounded on its east side by the Black Rock Ridge, and extending northward from the Boiling Spring to the Cañon Bluffs, at the north of Meadow Creek. This creek issues from the northern base of the Pillars of Atlas, and thence flows eastward through extensive meadows, till it empties into Upper Mud Lake, at the north extremity of the Black Rock Ridge.

The Great Boiling Spring is located at the southern extremity of this ridge, in latitude 40° 40′ north, and is about sixty miles distant from the west bend of the Humboldt River. The Lassen Trail passes by the spring, bearing in a north by east course up the valley, till it reaches Meadow Creek, where it makes a short deflection to the west, crossing a low divide to Cañon Valley, where it enters the High Rock Cañon, and thence passes northward.

East of the Black Rock Ridge, is Mud Lake Valley, which is a perfect desert, except where occupied by water ; between this valley and the Humboldt River, the country is entirely barren, and traversed by two low ridges, ranging north and south.

On a line with the Lassen Trail, the continuity of each ridge is broken transversely, whereby an easy grade is formed, presenting no impediment to the passage of a good road over them.

The above described district, from the Humboldt to the Sacramento River, through the Great Pass in the Sierra Nevada, was explored by me, in company with McNamey and Capt. Holloway, in the months of June and July, 1850, when emigrating to California. But the difficulties we encountered with Indians and want of suitable instruments, prevented us from making, in every particular, accurate observations. Nevertheless, as I have been governed by the rule to speak only of what I have seen, the descriptions here given of the relative positions of mountains and rivers, with the other important features of that region, may be relied on as generally correct.

As to the practicability of constructing a railroad through the above passes, I have not a shadow of doubt, and the more I examine the evidence derived from authentic sources, setting forth the merits of other passes, the more I am convinced that the above pass is greatly superior to any yet found in the Sierra Nevada or any other range of mountains on the continent of North America.

The great forest south and west of the pass, would furnish an unlimited supply of timber, even for a dozen railroads, while the fertility of the soil and the immense treasure concealed within the bosom of its placers, presents an extensive and inviting field to the industrious farmer and miner.

It is a well-known fact, to parties acquainted with the mountains, that there are many little valleys among them which are green with verdure, and wearing the smiles of perpetual spring while the adjoining mountains are covered with snow. Such, likewise, is the predominant feature of the valleys surrounding the Great Pass, and by reason of the low altitude of those valleys, and the manner in which they link together, snow will never form an obstacle to a passage through them.

The location of the Atlantic and Pacific Railroad is of vital importance to the American people,—especially those inhabiting Oregon and California, who expect to be mutually benefitted by its construction,—and this much desired object can only be accomplished by its introduction through a mediate passage to the Pacific Coast.

To this end, the above pass is admirably adapted : it being about equi-distant from the northern boundary of Washington Territory and the southern boundary of California, thus occupies a central position, which renders it equally accessible to the inhabitants of both districts ; and would, if chosen for the purpose aforementioned, avoid the necessity of bestowing partiality to one section, at the sacrifice of the other,—a circumstance that would certainly follow the selection of a more northern or southern route.

It is plain to any discerning mind, that a railroad crossing the continent by the northern borders of Nebraska and Washington Territories to the Pacific Coast, would be of little or no benefit to the inhabitants of California ; and, *vice versa*, a railroad crossing the southern borders of New Mexico to California, would confer no benefits to the inhabitants of Oregon or Washington Territories—especially those living in the interior parts ;—for the distance they would have to travel to reach the depot, is nearly equal to that direct to its terminus at the Missouri River.

This fact demonstrates the necessity of a central route, that parties both to the north and south may participate equally in its benefits ; and not only to them is it essential, but also to the parties owning the railroad : for, as from the branches of a tree the fruit is plucked, so likewise from the branches of the road the principal profits would be derived. Wherefore, then, establish a road to the extreme north, or south, where one side of the track is nearly all foreign ground, over which no branch can be laid ? And even if there was, it would be useless ; —as those regions are barren, and inhabited by the most hostile Indians in North America.

In locating a National Railroad, no sectional or speculative motive should be entertained ; but the great and all-absorbing object should be, to benefit the whole, and advance the common interest of an enlightened people.

To accomplish this grand object, the road should pass through the center of our dominions, by the most practicable route, and embrace the greatest range of usefulness possible.

A railroad commencing at any suitable point, near the Missouri River, and following the Platte River, and Sweetwater Valleys, to the South Pass, in the Rocky Mountains ; thence, by a descent, to the Bear River Valley, or otherwise following the elevated plateau, between Lewis River and Salt Lake Valley, till it reaches nearly to the Castle Rocks ; thence, by a southern deflection, and through the Thousand Spring Valley to the Humboldt River, down the valley of which, it would traverse to the west bend ; by leaving the river at this point and continuing westward, the road would pass between the Upper and Lower Mud Lakes to the Boiling Spring Valley, and cross said valley opposite the Great Boiling Spring to the Pillars of Atlas ; thence, passing between the Pillars it would enter McNamey Valley, up which it would continue in nearly a direct west course, to the Great Pass in the Sierra Nevada; thence, descending the western slope of said pass into Honey Lake Valley, crossing the same westward by the northern border of Honey Lake, to the eastern base of Pitt River Mountains; thence, crossing the summit of said mountains, by the northern base of the Black Butte, through Noble's Pass, it would then traverse over the Iron Hills in nearly a north-west course, to Lassen's Butte, and by following Captain Lyon's Trail around the north side of that Butte, would enter the Sacramento Valley at Battle Creek, without encountering through the whole route any more formidable obstacles than have been overcome in crossing the Alleghany Mountains by similar works.

The principal obstacles on the above line is met with at the South Pass and the northern spurs of Bear River, or Wah-Satch Mountains; the former, on account of its altitude and the snow, and the latter, on account of their ruggedness. But no fears need be entertained in regard to surmounting these obstacles as the ascent to the South Pass is so gradual, that a person would cross the summit without being aware of the act, were it not that the mountains north and south defined his position.

And all difficulties apprehended from heavy bodies of snow could easily be obviated—as at the summit of the pass, the only point where it would be likely to obstruct the road, would also be the point that would require excavating, in order to reduce it to a uniform grade: over the deep cut thus made, rough timbers can be cast, in the manner of rafters, put close together,—thus forming a substantial roof on which the snow could lodge, then the greater the quantity of snow, the better would be the shelter.

This method of making superficial tunnels could be followed at any point on the route where snow would be likely to form an impediment.

It is estimated that there is over 12,000 miles of railroads in the United States, at an aggregate cost of $356,000,000. The length of the Atlantic and Pacific Railroad would be nearly one-seventh of all the railroads in the United States, and at the same ratio of cost, would amount to about $50,000,000. But when we take in consideration the high price they had there to pay for land, the immense cost of timber, the great number of hills and dales required to be leveled, the many little streams that are bridged, and culverts made for the passage of common roads, we find that the ratio of their cost, exceeds that of the Atlantic and Pacific Railroad, which would lie mostly over extensive level plains, where the land is free, the timber free, but few streams to cross, and no culverts to be made for common roads to pass through.

It is estimated that the steepest grade of any railroad in the United States, is one hundred and sixteen feet to the mile; that of the Southwestern Tennessee is ninety feet to the mile, and the cars pass over it with perfect ease. The greatest amount of tunneling is on the Alexandria, London and Hampshire Road, in Virginia, being 19,536 feet, or over three and a-half miles. It is doubtful if there is any point on the Atlantic and Pacific route where the grade would exceed one hundred feet to the mile, and the aggregate amount of tunneling on the whole route, would probably exceed but little that of the one road abovementioned.

These facts show the possibility of establishing the railroad, and at a far less cost than heretofore estimated.

The line here indicated lies through the most interesting and valuable portion of Nebraska and Utah Territories, and is the only route yet designated which, throughout its entire length, with but few exceptions, offers inducements to settlers by being susceptible of agricultural development; while the temperate climate of the region through which it passes favors it greatly, when contrasted with the extreme cold of the more northern route, or the intense heat that pervades the extensive and terrible deserts lying further south.

These advantages, in connection with its central position, and its amplitude to subserve the great purposes designed in its construction, are plain and irrefutable facts of more weight than the wily arguments of speculators, or sectional advisers.

In the performance of so great a work as the formation of an Atlantic and Pacific Railroad, many unavoidable difficulties would have to be encountered on any route; but the benefits to be derived from the road are equally great, both nationally and individually. Its establishment is identified with the progress of this Republic, and would be a lasting monument of the enterprise and energy of its inhabitants.

For these reasons, and in view of the great interest involved, the above statements are respectfully submitted to your serious consideration.

I have the honor to be,

Very Respectfully,

Your Ob't Servant,

A. FREDONYER.

Hon. S. H. Marlette,

Surveyor-General of California.

www.ingramcontent.com/pod-product-compliance
Lightning Source LLC
LaVergne TN
LVHW020636110826
845149LV00004B/1217

* 9 7 8 1 4 1 8 1 9 1 0 7 8 *